Merrill, W. Margaret

Skeletons That Fit

DATE DUE			

About the Book

Do you know why there are mammals roaming the earth today, or what happened to the huge brontosaurus that lived millions of years ago? The answer has to do with bones.

Using clear and simple language, Margaret Merrill explains how each animal you see around you has a special bone structure that enables it to live in its environment. We go back millions of years to look at some of the ancestors of today's animals and learn why animals have bones; how bones, even mankind's, have changed over many years; and what may happen if an animal's bones don't change in this ever-changing world.

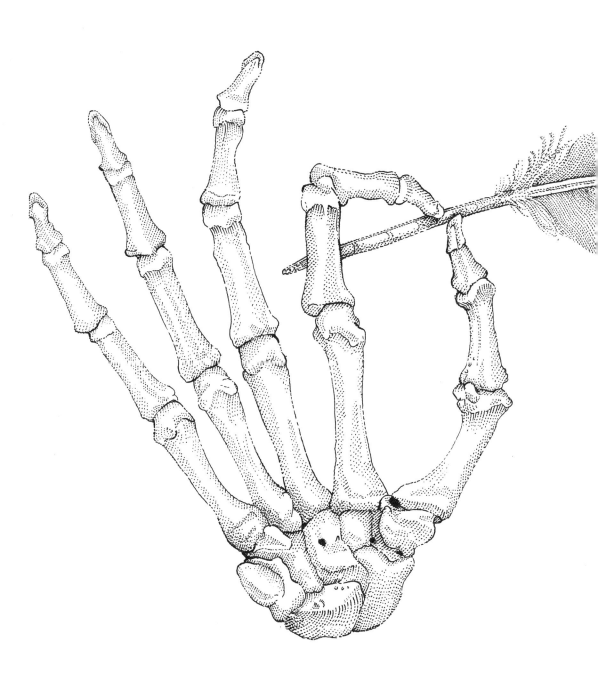

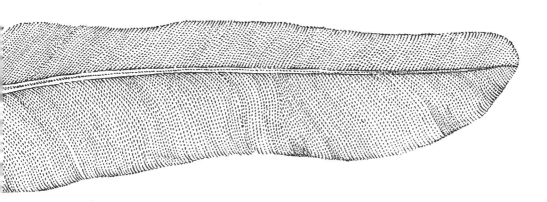

SKELETONS THAT FIT

by Margaret W. Merrill

illustrations by Pamela Carroll

COWARD, McCANN & GEOGHEGAN, INC.
New York

To Fred, Killy, and Gabe

General Editor: Margaret Farrington Bartlett

The author and publisher gratefully acknowledge the
assistance of Dr. Robert Chaffee, Director, Montshire
Museum of Natural Science, Hanover, New Hampshire,
and Earl Manning, Department of Vertebrate
Paleontology, The American Museum of Natural
History, New York, New York.

Merrill, Margaret W.   Skeletons that fit
Bibliography: p.
Includes index.
SUMMARY: Surveys the evolution of animals according to
their skeleton.
1. Evolution—Juvenile literature.   2. Skeleton—Juvenile
literature.   3. Paleontology—Juvenile literature.
[1. Evolution   2. Skeleton]   I. Carroll, Pamela.
II. Title.
QH367.1.M47 596′.03′8 77-24155

Contents

Clues from the Past   7

No Bones About It: The Invertebrates   13

The Fish That Walked   19

Amphibians in Between   25

Creatures on a Larger Scale: Reptiles   29

The Lizards That Flew   37

Warm and Furry: Mammals   41

Gone, Except for Their Bones   47

Vertebrate Chronology   57

Glossary   58

Books for Further Reading   60

Index   62

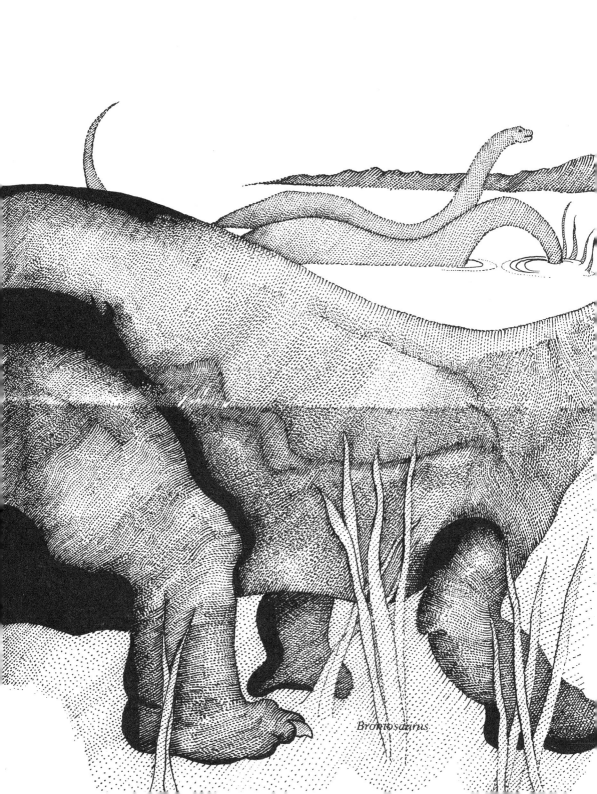

Brontosaurus

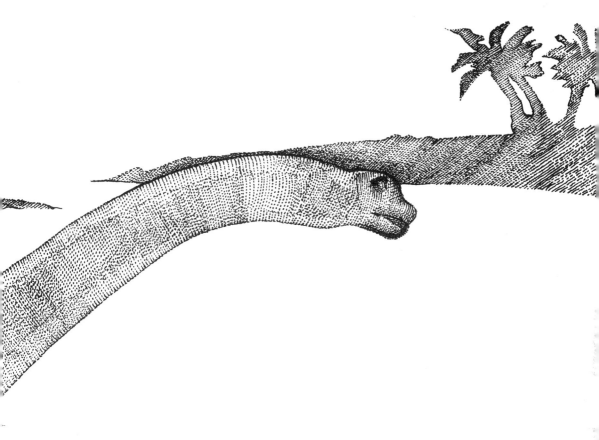

Clues from the Past

Have you ever wanted to go back far into the past to see the strange animals, to figure out what they looked like? You can, even today. You don't need a time machine, but you do need some clues.

If you are in the country you can look for clues in a dry riverbed or deep in a limestone cave. Or ask your favorite older person to look with you in mines or quarries deep beneath the earth's surface. There, you may find fossil bones.

Animals that have bones need them for support and shape, for movement, and for protection.

7

An animal's bone structure—the skeleton—supports its body and gives it a shape—tall and wide, short and narrow, and so on.

Bones also help the animal move. They are levers for the muscles to work on. The muscles pull the bones up, down, or sideways like strings pull a puppet's wooden arms. Muscles and bones must work together to help an animal move efficiently. The shape of a bone is a clue that shows whether an animal moved best in water, on land, or in the air.

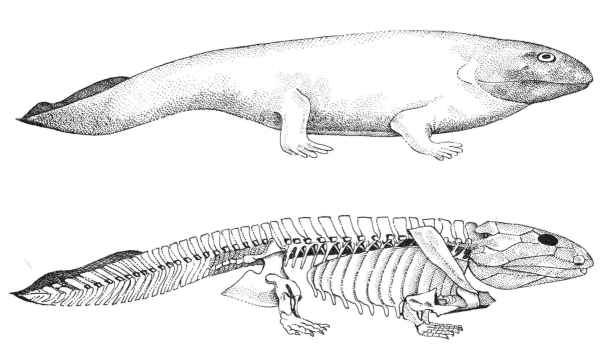

This is one of the earliest known amphibians. Its tail was still fishlike. See the small fin on top? Its limbs were short, not much longer than fins, but they did make it far easier for the amphibian to get about on land.

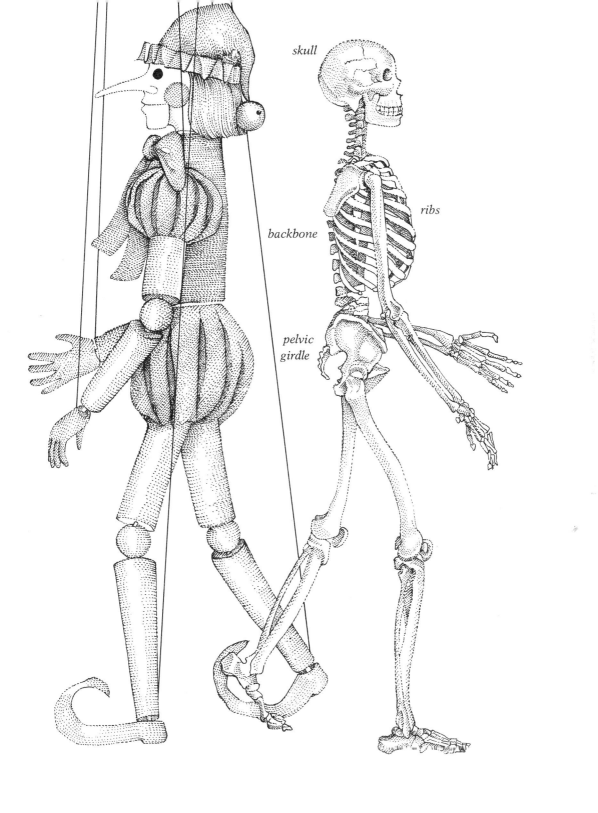

skull

ribs

backbone

pelvic
girdle

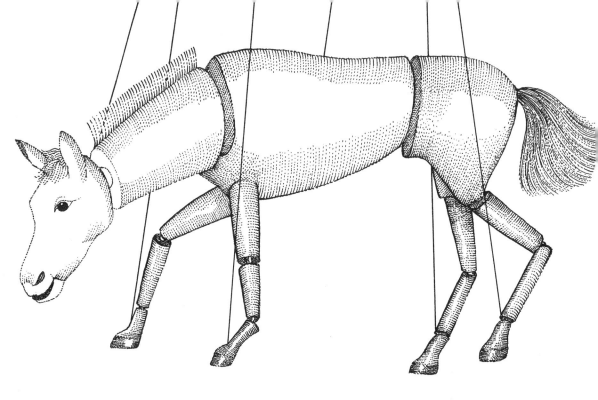

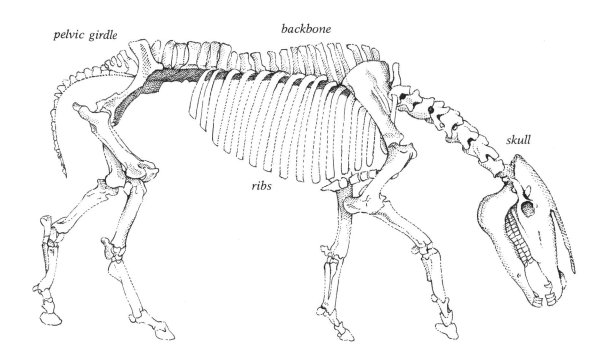

pelvic girdle

backbone

skull

ribs

Bones also protect the soft parts of an animal's body. An important bone, the skull, covers the brain and forms bony protective ridges above the eyes. Attached to the skull is the backbone, which works like a shock absorber. The backbone cushions the skull, protecting it from hard bumps. Ribs join the backbone and circle the chest, giving protection to the lungs and heart. Some animals have a heavy pelvic girdle. This pelvic girdle protects the stomach while the animal hunts for something to put in it. It is a cradle for some animals' young, a protected place to grow in, before they are born.

Knowing that bones are needed for support, for shape, for movement, and for protection can help when you find fossil bones. Each bone is a clue to the animal's way of life, for each bone has adapted to fit the needs of a special environment.

If you go to a museum you may see how fossil bones are put together to make a skeleton. When an interested person finds enough bones, he or she can tell how large an animal was, when and where it lived, and what it ate.

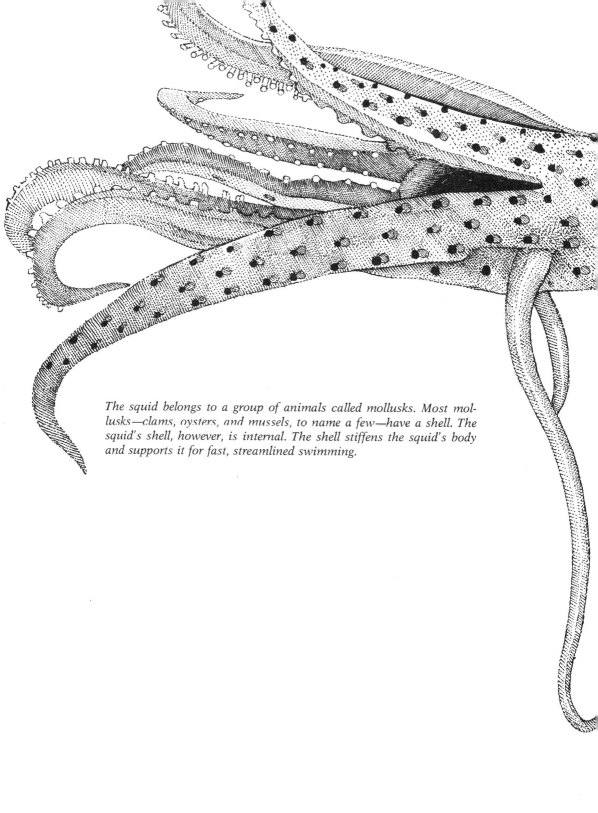

The squid belongs to a group of animals called mollusks. Most mollusks—clams, oysters, and mussels, to name a few—have a shell. The squid's shell, however, is internal. The shell stiffens the squid's body and supports it for fast, streamlined swimming.

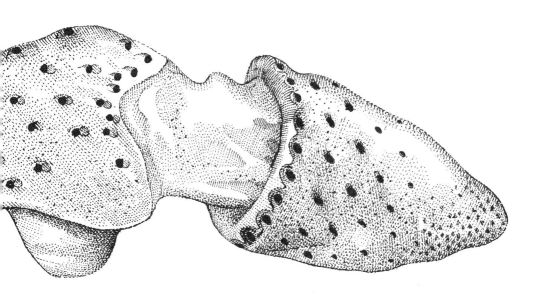

No Bones About It: The Invertebrates

Animals that have bones could not get along without them. But there was a time long ago when no animals had bones.

The earth is roughly 5,000 million years old. The first life forms appeared on earth about 1,500 million years ago. These first life forms were bacteria, followed later by simple plants, such as algae.

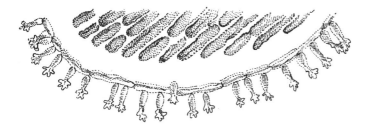

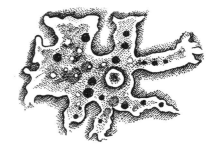

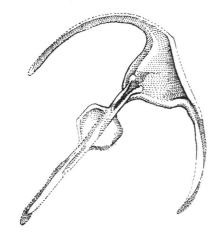

These microscopic animals belong to a group called protozoa, which means first animals. To the naked eye, the largest protozoa looks like a white speck. There are thousands of different protozoa alive today. More of them exist than of all other animals combined!

About 1,000 million years ago, the earth's first animals appeared in the sea. These animals had no bones. They looked so much like the simple plants that you could hardly tell which was which. The sea was like a watery cushion. These animals didn't need bones for protection; the sea surrounded their delicate bodies. These animals didn't need bones for movement; sea currents swirled them about. They rode the waves like es-

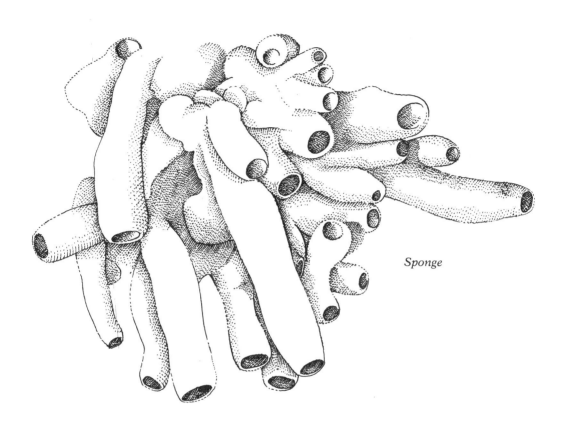

Sponge

calators in search of food. Because they had no bones for structure, these unicellular animals were very small and shapeless.

For some of these animals the no-bones system worked fine, and many of their look-alikes are floating in our oceans and ponds right now. Of these, the sponges, worms, and jellyfish may be familiar to you. Many other boneless animals appear in your microscope when you are looking at a drop of pond water.　15

Animals like clams, lobsters, sea urchins, centipedes, and mosquitoes also have no bones.

But some do have shells or brittle skin to protect their bodies. A shell is an outside skeleton (called an exoskeleton). But shells are not bones. Shells are made of a fairly simple and common chemical, calcium carbonate. Bones are made of a much more complex and rare chemical, calcium phosphate, and a gelatin which makes them slightly flexible. Besides being inside the animal, bones make red blood cells and store calcium.

All animals without bones—from sponges to jellyfish to lobsters—are called invertebrates, which means "animals without a backbone."

Shells work like bones to protect the animal, although they are an outside protector.

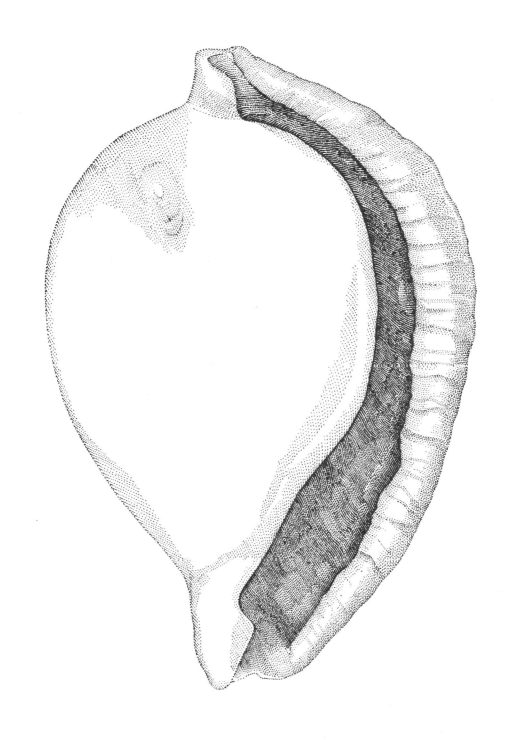

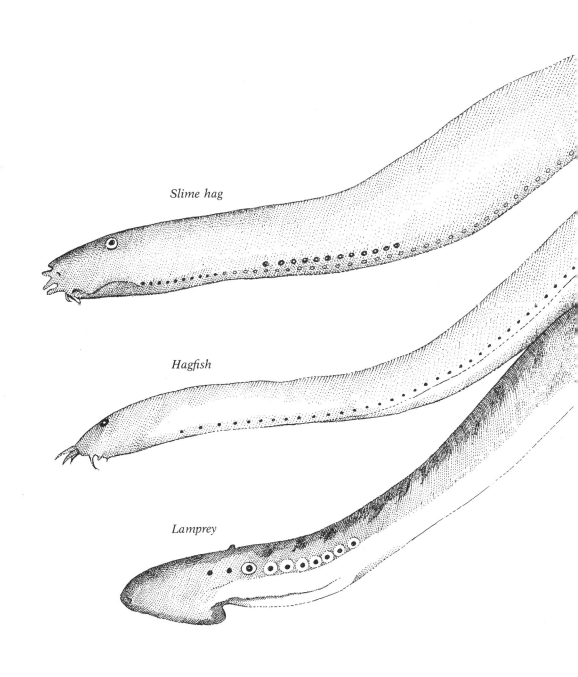

Slime hag

Hagfish

Lamprey

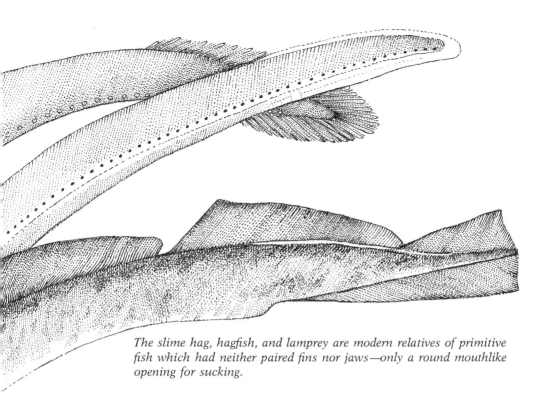

The slime hag, hagfish, and lamprey are modern relatives of primitive fish which had neither paired fins nor jaws—only a round mouthlike opening for sucking.

The Fish That Walked

Back in that swirling sea 1,000 million years ago, some invertebrates began to change into other forms of life. No one knows exactly why or how an animal's body begins to change, but it may have to do this to fit into a new or changing environment.

Over many years some invertebrates were carried to another part of the ocean. There, they had to move faster and catch new food. The animals whose bodies swam fastest and caught the most food lived. The others died out or moved away or changed into other forms of life. 19

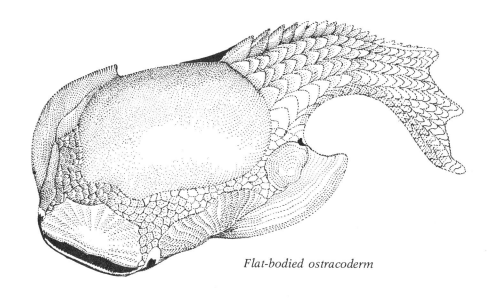

Flat-bodied ostracoderm

When animals' bodies change, they change very slowly. Often it takes many generations for one group of animals to look even slightly different. Five hundred million years went by during which time some animals began to look a little bit "fishy."

The first real fish, the ostracoderm, had an exoskeleton like a lobster, but this skeleton was made of bone, not shell.

Later, fish developed a backbone inside. With a snap of the spine, these fish shot through the water toward their prey. This was a great improvement over floating, bobbing, or paddling after food. Besides being very speedy, fish had another advantage over their prey—they had bony jaws to eat it with.

Still later, many other kinds of fish developed. Their bodies had changed to adapt to new parts of the ocean. But there were two main kinds of fish—you could tell by their fins. The fins of one group—called the ray-finned fish—were flat and bony. The fins of the other group were rounded out by flesh and muscle, and by short thick bones. These fish were called lobe-finned fish because lobe means rounded. The few living relatives of the lobe-fins, the lung-fish and the coelacanths, are very important to scientists. They study them because the bones of the lobe-fins went through many changes. The lobe-finned fish is ancestor to all the creatures with a backbone that came to live on land.

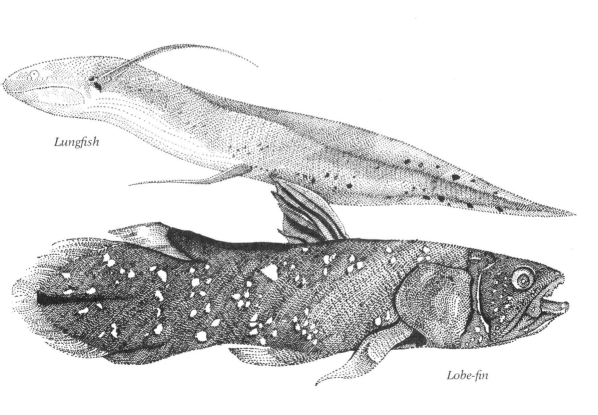

Lungfish

Lobe-fin

The lobe-fins once swam in shallow bays and river mouths. At first, this was a new environment. Space and food were plentiful.

When the shores became crowded, some went to live in fresh water. This was yet another new environment for animals.

Lobe-finned fish had lungs and were able to breathe through the mud in seasons when their ponds dried out. One hundred sixty million years went by. Some lobe-fins changed as they searched for new places to live. The bones in their fins grew longer and thicker. At the tip, the bone separated and became like five short toes. The fins had become the legs and feet of a new animal.

Did the first animal to crawl out of water and onto land look like this?

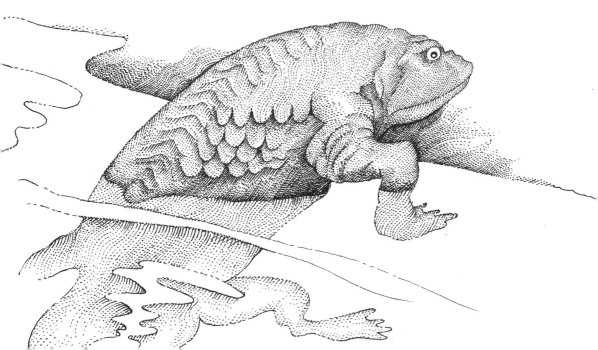

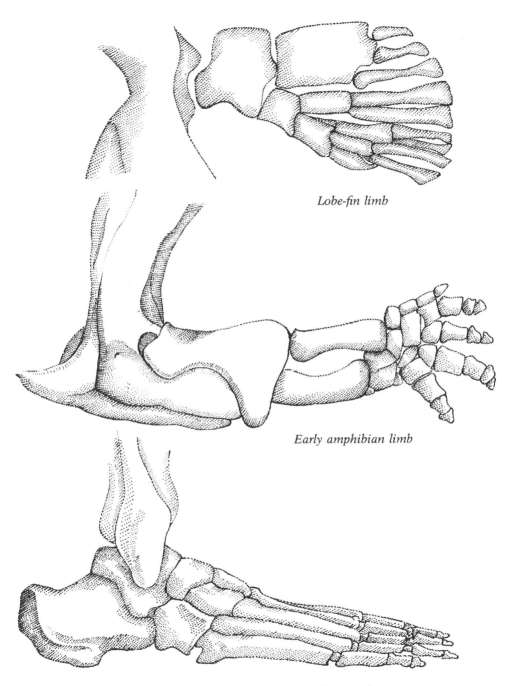

Lobe-fin limb

Early amphibian limb

Human foot

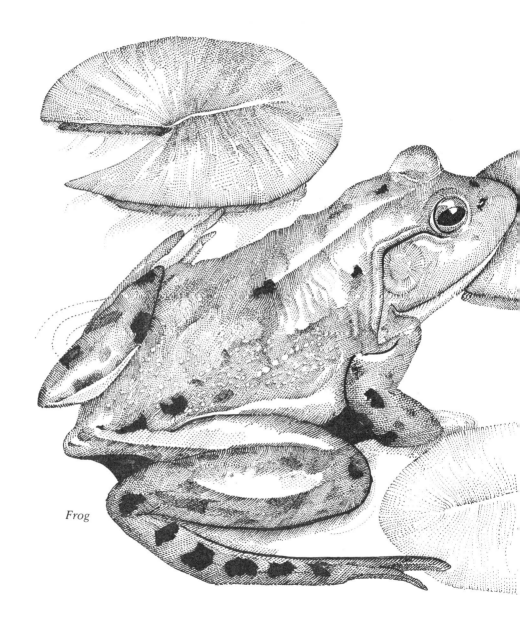

Frog

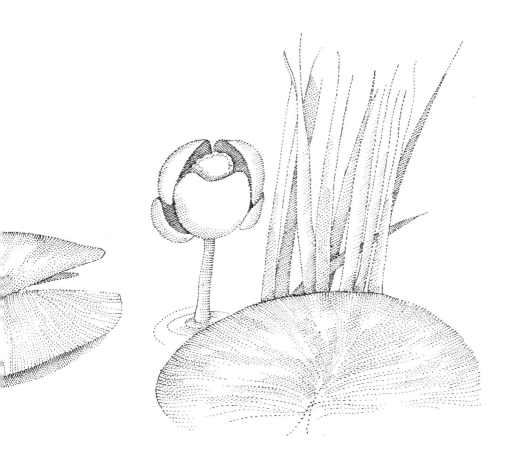

Amphibians in Between

This new animal spent most of its time in the water as a fish. Perhaps its trips on land were to search for a new swimming hole when the old one became crowded or dry. The animal was an amphibian, which means adapted for life in the water or on land. You can find most amphibians today always in or near water. They are salamanders, toads, and frogs. Some clues for finding them are the jellylike eggs you see floating in ponds in the springtime. They belong to these animals, for most amphibians must lay their eggs in water.

It was the same for the early amphibians. Often the water was not a safe place for the eggs. There were many kinds of fish searching for food 360 million years ago. The amphibian eggs made a delicious meal.

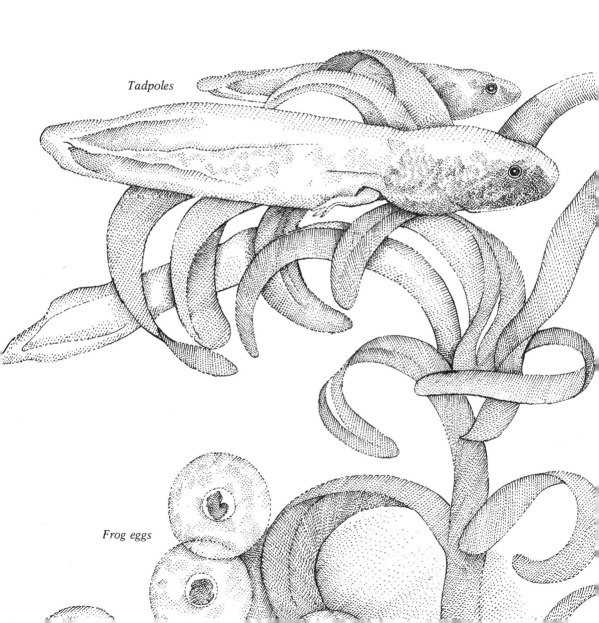

Tadpoles

Frog eggs

Young frog

Land was a much safer place. Only a few primitive plants, insects, and shelled creatures lived there. The eggs and bodies of some amphibians began slowly to change for life on land. Some kinds of amphibian eggs slowly developed shells. Again no one knows how, but some amphibian bodies became protected by specialized scales. These scales held moisture in an animal's body when it strayed far from swamps and streams. When other changes in body and bone structure were complete, the new animal was called a reptile. Reptiles appeared about 310 million years ago.

Young alligator

Adult alligator

Creatures on a Larger Scale: Reptiles

Scales covering its body and the eggs which it laid on land made the early reptile different from an amphibian, but their bones were very much alike. Mostly, the reptile's skeleton had changed to make it taller and better coordinated on land.

To see how the two are different, let's look at a salamander (amphibian) and an alligator (reptile). If you can, catch a salamander and set it down near its pond. Its legs are so short and so extended to the sides that it rests on its belly. It looks as if it is swimming when it walks. Its whole body twists before it can move its head because amphibians have practically no neck bones. There! It's back in the pond already.

Now for the alligator. Perhaps you can see one at a zoo. Compared to the salamander, the alligator is a real ballerina. When it wants to, an alligator can walk high off the ground. Powerful legs come straight up underneath it. An alligator walks gracefully when it wants to move, but you have to watch carefully, for most alligators would rather bask in the sun. It doesn't have to move all around to see, as a salamander does. The alligator has more bones and longer bones in its neck than the amphibian salamander. These neck bones make it easier for a reptile to turn its head without moving its body.

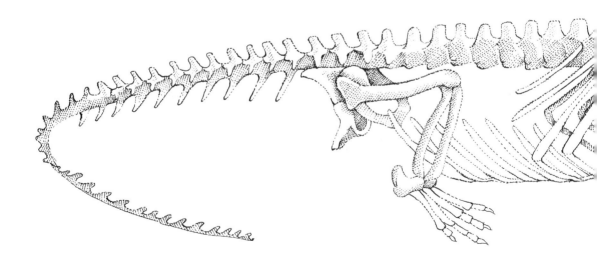

Salamander

As time went on, reptiles began to look very different from one another. Their bones began to come in assorted sizes and shapes. As food became scarce in some hunting grounds, some reptiles went to live in the sea. Some slithered under rocks. Others grew so large that they frightened the more timid reptiles away. But, huge as they were, many of these reptiles became extinct. Their bones could not adapt to the demands of a changing environment.

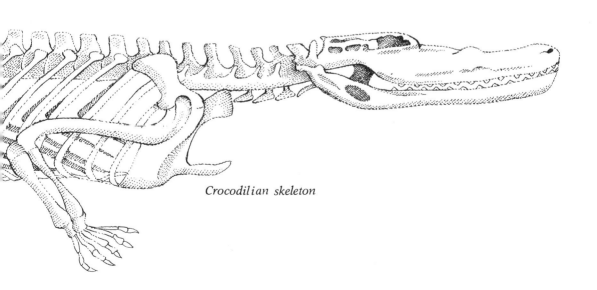

Crocodilian skeleton

One such reptile was the brontosaurus, a dinosaur. Brontosaurus bones were huge. It lived in water, where at least part of its enormous bulk could be floated about. It had many long bones in its long, long neck and could reach deep under water for plant food. But when the environment changed, the brontosaurus died out.

The bones of other reptiles were smaller, and these animals fitted more easily into new environments. One reptile which has long outlasted the dinosaur is the turtle.

The turtle has a skeleton both inside and out. The exoskeleton is the turtle's shell, which is made of bone and grows with the animal. It protects the turtle well. When a turtle wants to hide, the bones in its neck pull its head under the shell to safety. This is one retreat that wins the battle!

Turtles have a skeleton inside and out—the shell is made of bone, too!
Turtles are the oldest living group of reptiles on earth today.

Unlike many animals, reptiles are never fully grown. Their bones grow during the reptile's entire life.

The snake, a very successful reptile, goes to great lengths to prove this. Boas 20 feet long slither through the jungles of South America. To watch it move, you'd think the boa had no bones at all. Powerful muscles contract, and scales on its belly push it forward. The snake's skeleton is streamlined, so streamlined that it doesn't have any legs.

Sometimes you may see tiny claws on a snake's body. These mark the spots where snakes once had legs, many thousands of years ago. Instead of legs, today's snake has a special long backbone made up of many joints. These are called ball and socket joints because the ball of one joint fits into the socket of the next joint like beads on a pop-it necklace. The joints rotate so that the snake can twist or coil.

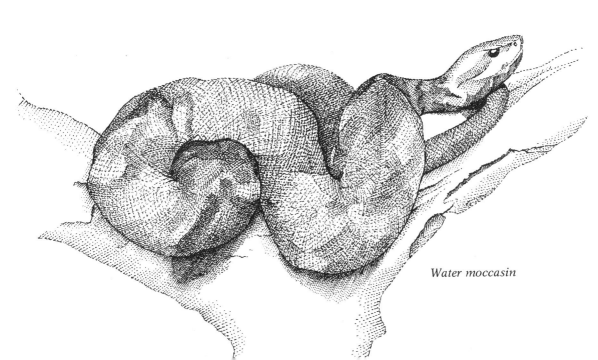

Water moccasin

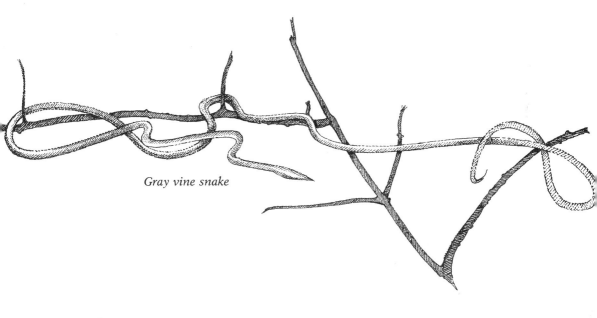

Gray vine snake

The snake is the only animal that can change shape to fit its house, whether a crack in a stone wall, a sandy burrow, or a hammock made of itself looped over a jungle branch. Because of its adaptable shape, the snake, too, has been around a long time. Snakes, turtles, crocodiles, and lizards—some of today's reptiles—are in many ways like their relatives of millions of years ago.

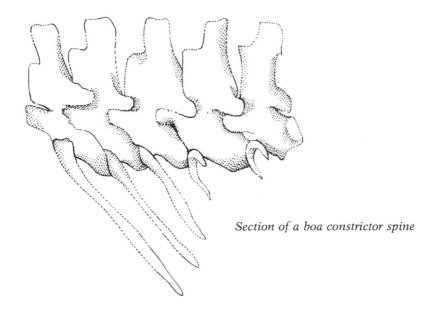

Section of a boa constrictor spine

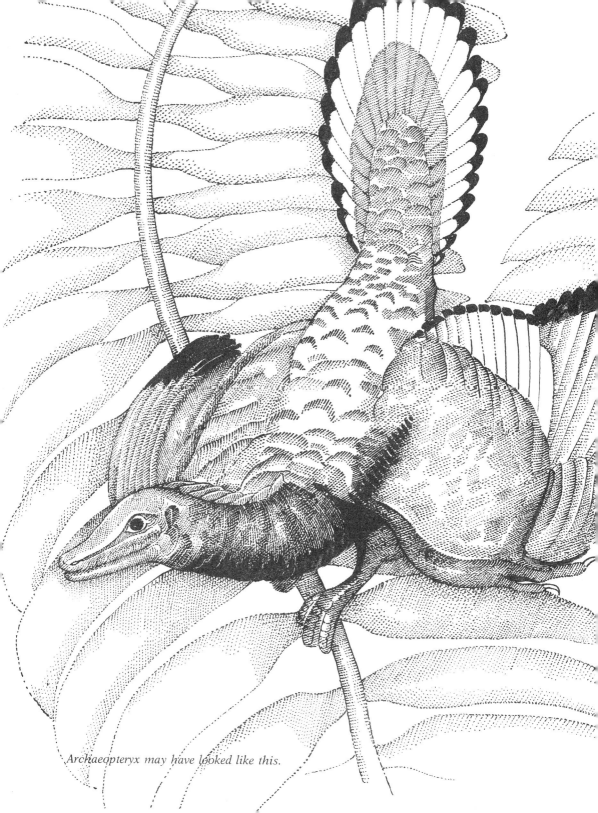

Archaeopteryx may have looked like this.

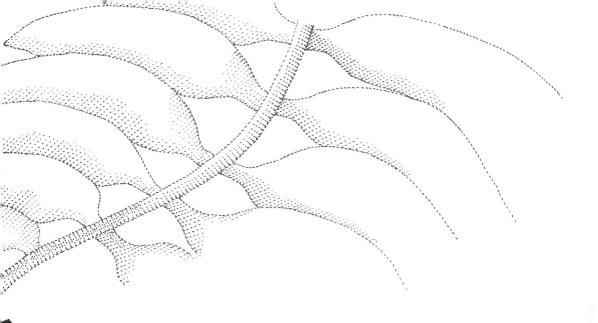

The Lizards That Flew

Not all the reptiles that lived millions of years ago became
snakes, turtles, crocodiles, or lizards. Not all reptiles kept on
being reptiles! In some forests food became scarce on the
ground. There was much competition for space to live on the
forest floor. Some reptiles moved into low trees. These provided
a less crowded environment. Food was more plentiful and it
was safer there.

In at least one forest some lizards had become good jumpers
and gliders. But now some of these high fliers developed feathers
instead of scales. Scientists have found clues that suggest what
these creatures may have looked like. Two fossil skeletons show
that the lizard-bird was about the size of a crow. Prints of feathers showed clearly that here was a new animal. Scientists
named it Archaeopteryx, which means "old bird." 37

Birds have changed a lot since archaeopteryx, 150 million years ago. No one knows how birds discovered the safety of flight. No one knows how their bones began to change as they began to fly farther. Today birds' bones are hollow, and most birds are truly airborne. The hollow bones are braced inside by thin strips of bone which run crosswise like the struts in a plane's wing. Some of the bird's bones are fused or joined together like the pieces of a balsa glider when it is ready for flight. This strengthens the bird's frame and combats wind pressure when it flies.

Have you ever held a bird in your hand? It feels like nothing is there. A scientist at the American Museum of Natural History once weighed the skeleton of a frigate bird. Frigate birds chase other birds and steal their food, so they must be quick and powerful in flight. They have a wingspan of 7 feet, but a frigate bird's bones weigh only 4 ounces, or about 100 grams. Their entire skeleton is lighter than their feathers!

Frigate bird

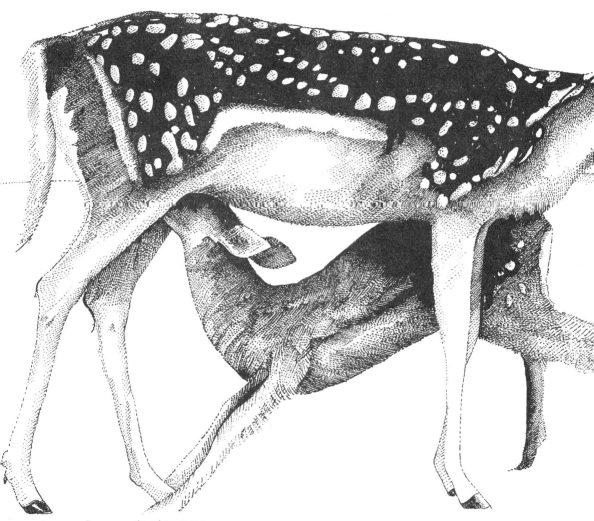

Deer nursing its young

Warm and Furry: Mammals

Just before the time the lizard-birds were hopping in the ancient forest (about 200 million years ago) a new animal developed. You would never have noticed this creature if you had been alive then, for it was still the Age of the Great Dinosaurs. This new animal was about the size of a chipmunk. Like archaeopteryx, it developed from a reptile. Its bones were like a reptile's bones, but there was an important difference. This new animal nursed its young, as bears, elephants, monkeys, and people do today. It was called a mammal.

Some mammals, and in particular the great apes which walk upright, have thick strong bones in the pelvic girdle to hold the young before they are born.

The warm-blooded, and usually furry, mammal was very successful. There are thousands of kinds of mammals alive today in every part of the world.

Why were they so successful? Even though the rest of its bones were about the same size as those of its closest reptile relative, the early mammal's skull grew larger with time. The mammal's brain had more room to store information. Because of their more complex brain, mammals can learn through memory and training. They can protect, guide, or teach their young.

The gorilla—a great ape—cares for and teaches its young until it is nearly three years old.

There were important differences, too, between mammal and reptile bone structure. Remember the alligator? Though it can walk gracefully on land, its backbone is fixed. It has an arch in it which will always be there; but a cat, which is a mammal, can arch its back any time.

Domestic cats

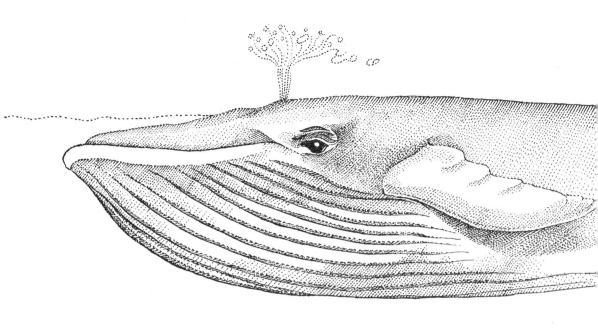

A mammal's backbone bent like an elastic bow. Mammals' leg bones were longer too, and these, with the bendable backbone, increased their running speed. If a mammal couldn't outwit its enemies, it might be able to outrun them!

Because of their warm blood, mammals were able to live in climates that would have sent reptiles into year-round hibernation. Today there are walruses in the Arctic, caribou on the tundra, whales in the sea, and moles, which if they are anywhere else, would rather be underground.

But these animals took a long time getting fit for their new environments. Today, moles have powerful bones in their shoulders and feet for digging. But the generations of animals on the way to becoming moles may have looked more like mice or like tiny creatures we've never seen.

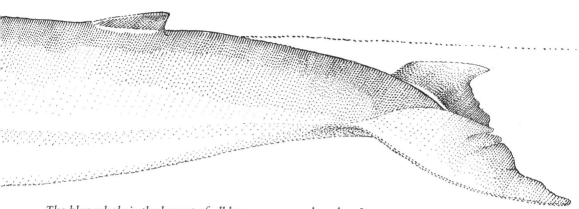

The blue whale is the largest of all known mammals today. It can grow up to 100 feet long.

Look at the whale, for instance. Like all mammals it once had a nose with nostrils at the tip. But the whale was an unusual mammal. For millions of years mammal skeletons had adapted for life on land. Then the whale—or its ancestor—returned to the sea to live like a fish. No one knows why the whale left the land or what it looked like before it went to sea. Perhaps, like all the animals who had gone to live in a new place, the whale was hungry. When it moved to the ocean it needed a better way to breathe. In order for the whale's nostrils to become a blow-hole, the skull had to change. Anyone who jumps into the sea does not automatically get a blowhole. Scientists believe that the nostrils of whales took many, many generations to shift slowly to the top of the skull. Because of the blowhole on the top, a whale does not need to lift its heavy head out of the water to breathe.

The giant ground sloth may have looked like this.

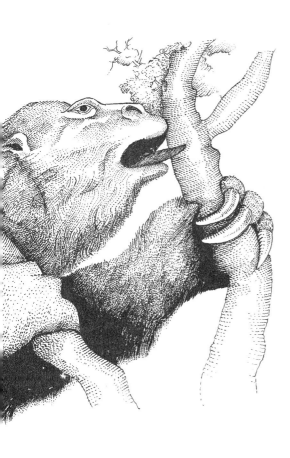

Gone, Except for Their Bones

All the animals alive today are only a small part of all the animals that have ever lived. Not all the animals of the past were
as lucky as the whale. For some, the environment changed too
quickly for their bodies to fit in. One animal whose bones take
up a lot of space in museums is the giant ground sloth. This
sloth was twice as large as a grizzly bear. The sloth's skull was
small, and its simple thought-patterns could not help it in a
new situation. When new enemies, the saber-tooth tiger and
tribes of early man, hunted it, the sloth was too slow to escape.   47

Irish elk

Another animal that is gone now must have been very handsome. The leg bones of the Irish elk were so long that you would have been able to walk—almost standing up—under its belly. The antlers of the Irish elk were larger and thicker than any of its bones. Twice the size and weight of a moose's antlers, they spread 11 feet from tip to tip. Perhaps the antlers were an attempt to equip the elk for defense against wolves, but it was an attempt that backfired. The antlers made the magnificent elk top-heavy. Scientists believe that the elk's body became weakened over generations when it had to give up so much energy to support and grow the huge structure. Too, the elk became tangled in the brush and many lay trapped in the peat bogs of Europe. Gradually the herds dwindled.

A third animal, a bird called a moa, also could not escape when its environment changed suddenly. Moas lived on the island of New Zealand. They were bigger than ostriches, almost 10½ feet tall. The rest of the animals on New Zealand were smaller vegetarians. They had never tried moa meat. The moas, too, found plenty of food on the ground. Over the years moas stopped flying, and their wir.g muscles and bones weakened so they could not fly at all. About 600 years ago a tribe of humans, the Maoris, settled on the island. They hunted the moa and its eggs for food. Like countless birds and animals since, the moas were completely destroyed by man.

Moas

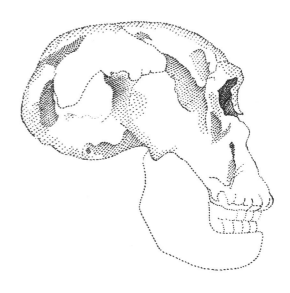

Homo habilis

Man, the latecomer, is one of the most remarkable of animals. Man's bones have changed a lot since the early men who appeared in Africa, possibly 3 million years ago. Though fossil remains are rare, scientists have found fragments of the early man's skull which show that his brain was about half the size of the brain of present-day man, *Homo sapiens*. Early man's brain was slightly larger than the ape-man's who lived at the same time—and he used tools, crude pieces of chipped quartz. Scientists named him *Homo habilis*, or handy man.

More clues about man's ancestors were found in a quarry in Peking, China. There, deep in the limestone, where caves once were, scientists discovered fragments of other human skulls. They pieced these fragments together and measured them

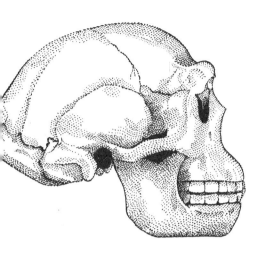

Homo erectus

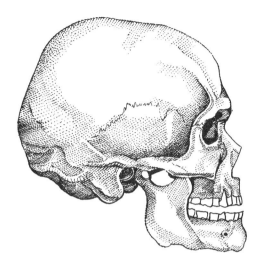

Homo sapiens

to discover that this man's brain capacity was slightly larger than the brain capacity of *Homo habilis*—about two thirds the size of the brain of present-day man. A mold was made of the inside of the skull bones. The pattern of folds of the mold showed that Peking man was capable of language. A variety of tools found told of his ability to reason and create. Straight leg bones were clues that Peking man walked erect. Peking man is also called *Homo erectus*, which means straight-standing man.

Since *Homo habilis*, the human brain has changed at a phenomenal rate. In less than 3 million years it has almost doubled in size. When you look at the Vertebrate Chronology chart you can see that this is a very rapid change in a history of long, slow changes.

51

With increased brain size and fingers that can pick up a feather, humans have become very clever, capable of an ever-widening range of tasks. Today, when we explore a new environment, we do not grow a new structure such as a fin, a foot, or a wing. We build it. Today our boats go out on the sea, our planes and rockets fly into space, and our vehicles travel on land almost everywhere.

People and the things we have built take up a lot of the earth's space. Much of this space—open prairie, forest, sea marsh, tundra, jungle, and other environments—also belongs to other mammals, to birds, to reptiles, to amphibians, to fish, and even to the tiniest invertebrates.

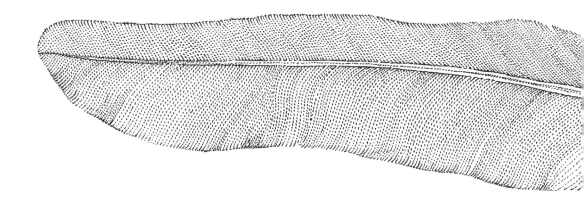

Now, as we find new clues, we see that many of our ancient bony relatives such as the dinosaur and the Irish elk had trouble fitting into a changing world. And we see that by means of the special bone structures which took millions of years to develop, today's animals have skeletons that fit each to its own particular space. These bone structures are not like the structures people build in a few months—or days. The skeleton which an animal needs for support and shape, for movement, and for protection is a unique and irreplaceable part of today's changing world.

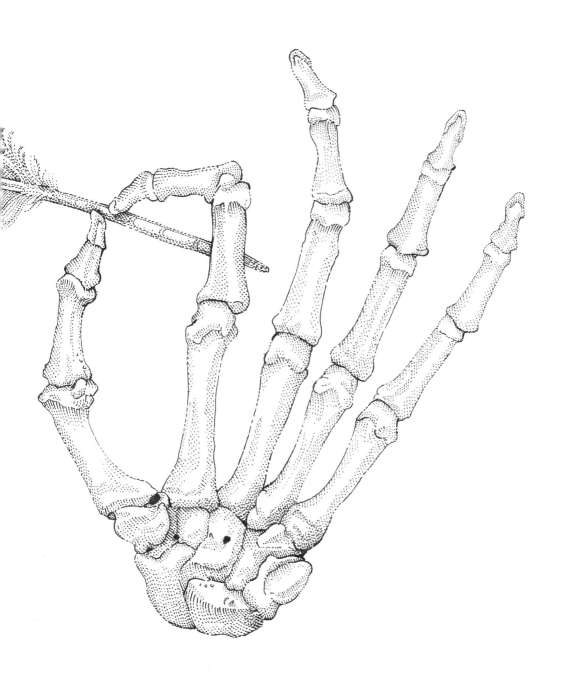

Vertebrate Chronology

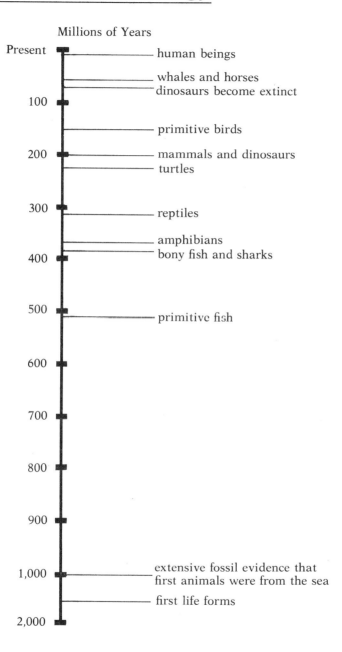

Millions of Years

Present — human beings

— whales and horses
— dinosaurs become extinct

100

— primitive birds

200 — mammals and dinosaurs
— turtles

300 — reptiles

— amphibians
400 — bony fish and sharks

500 — primitive fish

600

700

800

900

1,000 — extensive fossil evidence that
first animals were from the sea

— first life forms

2,000

Glossary

ADAPT   To make fit for a new use or a new situation.

AMPHIBIAN   An animal which is adapted for life in the water and on the land.

BIRD   A warm-blooded, feathered, egg-laying vertebrate whose forelimbs—wings—have adapted for flying.

BLOWHOLE   The nostril of seagoing mammals such as whales, porpoises, and dolphins, which has moved to the top of the head.

CAPACITY   How much something can hold, store, or contain.

COMPLEX   An organ or organism which is made up of many parts.

ENVIRONMENT   Surroundings, climate, and places in which an animal lives.

EXOSKELETON   An outside covering or structure which protects the animal, such as the lobster's shell.

EXTINCT   An animal or plant which no longer exists.

FOSSIL   A part or a tracing of an animal or plant of the past which has been saved inside the earth's crust, or any evidence of prehistoric life.

HIBERNATE To pass the winter in a sleep state, with life processes functioning on a low level. No growth occurs during hibernation.

INVERTEBRATE An animal without a spinal column.

LOBE-FINNED FISH A fish whose fins contain thick, short bones and a heavy muscular base.

MAMMAL A warm-blooded, usually furry, animal which nurses its young.

PELVIC GIRDLE The part of the skeleton which attaches to the vertebral column and from which the hind limbs (in man, the legs) are supported.

PROTOZOA A highly varied group of microscopic, single-celled animals.

RAY-FINNED FISH Fish whose fins are flat and are made of a web of skin stretched over long straight (raylike) spines.

REPTILE A cold-blooded, scaly animal whose young, with a few exceptions, hatch from eggs.

SPINAL CORD A collection of nerve fibers which is enclosed within the vertebral column or backbone.

UNICELLULAR A plant or animal which has or consists of only one cell.

VERTEBRATE Any animal such as a fish, bird, or mammal which has a backbone or spinal column.

Books for Further Reading

ANDREWS, ROY CHAPMAN, *All About Strange Beasts of the Past.* New York: Random House, 1956. Roy Andrews, former director of the American Museum of Natural Science, tells of his exciting discoveries of fossils of animals never seen by man, the hornless rhinoceros which was 25 feet tall, the shovel-jawed mastodon whose lower jaw was almost as long as the animal was tall, and many other strange beasts.

CLYMER, ELEANOR, *Search for a Living Fossil: The Story of the Coelacanth.* New York: Holt, Rinehart and Winston, 1963. A boat fishing off the east coast of Africa pulls in a fish with steel blue scales—a fish so strange that the old fisherman who saw it lying on the deck didn't think it was a fish at all. The coelacanth, called a living fossil by scientists who thought it extinct for 70 million years, turns out to be a relative of the lobe-finned fish, the ancestor of all animals with a backbone.

FACKLAM, MARGERY, *Frozen Snakes and Dinosaur Bones: Exploring a Natural History Museum.* New York: Harcourt, Brace, Jovanovich, 1976. A behind-the-scenes tour of what goes on in a natural history museum. In the delightful text, Ms. Facklam describes museum operations from identifying and cataloging new finds to reconstructing the stegosaurus from its bones.

RAVIELLI, ANTHONY, *From Fins to Hands: An Adventure in Evolution.* New York: Viking Press, 1968. With lively illustrations accompanying the text, Anthony Ravielli, noted contributor to *Sports Illustrated,* traces the evolution of the hand from the bone structure of the fin of the early lobe-finned fish to the development of flexible hands with movable fingers in the primates, the ancestors of man. Last comes man with his "dextrous hands controlled by a thinking brain."

ROMER, ALFRED SHERWOOD, *The Vertebrate Story,* 4th edition. Chicago: The University of Chicago Press, 1959. This book is an account of the development of the back-boned animal from the first highly varied types of fish to the family of man. It outlines the animal's biological structure, its function, and its ways of life with the evolutionary story as the leading theme. Recommended for students and other interested adults.

SPINAR, Z. V., *Life Before Man.* New York: American Heritage Press (McGraw-Hill, Inc.), 1972. Simple divisions of time before man show the life forms which flourished in each era. Pictured is the earth's beginning, imaginatively reconstructed, the development of the first plants, the development of the first animals without a backbone, and later the fish and other creatures of the sea. There is a large section on the Age of the Ruling Reptiles, including illustrations of the dinosaurs and winged reptiles.

Index

Italic numbers indicate illustrations

Alligators, *28*, *29*, 29–30, 43
Amphibians, *8*, *23*, 25–27, 29–30
Antlers, 48
Apes, 42
Archaeopteryx, *36*, 37

Backbones, *9–10*, 11, 20, 34, 43, 44
Bacteria, 13
Birds, 37–39, 49
Blowhole, 45
Blue whale, *44–45*
Boa constrictor, 34, *35*
Brain, 42, 50–52
Brontosaurus, 32

Caribou, 44
Cats, 43, *43*
Centipedes, 16
Clams, *12*, 16
Coelacanths, 21
Crocodilian, *30–31*

Deer, *40–41*
Dinosaur, 32

Elk, 48
Exoskeleton, 16, 32

Fins, *8*, 21, 22
Fish, 19–22
Foot, *23*
Fossil bones, 7, 11, 50
Frigate bird, 38, *39*
Frog, *24–25*, 25, 27
Frog eggs, *26*

Gorilla, *42*
Gray vine snake, *35*
Ground sloth, *46–47*, 47

Hagfish, *18–19*
Homo erectus, *51*, 51
Homo habilis, *50*, 50, 51
Homo sapiens, 50, *51*

Invertebrates, 13–16
Irish elk, *48*, 48

Jellyfish, 15, 16

Lamprey, *18–19*
Lizard-bird, 37
Lobe-finned fish, 21, *21*, 22, *23*
Lobsters, 16
Lung-fish, 21, *21*

Mammals, 41–45
Man, 50–52
Moa, 49, *49*
Moles, 44
Mollusks, *12*
Mosquitoes, 16
Muscles, 8
Mussels, *12*

Ostracoderm, 20, *20*
Oysters, *12*

Peking man, 50–51
Pelvic girdle, *9–10*, 11, 42
Protozoa, *14*

Ray-finned fish, 21
Reptiles, 27, 29–35
Ribs, *9–10*, 11

Salamanders, 25, 29–30, *31*
Scales, 27, 29
Sea urchins, 16
Shells, *12*, 16, *16*
Skulls, *9–10*, 11
Slime hag, *18–19*
Sloth, 47
Snakes, 34–35
Sponges, *15*, 15, 16
Squid, *12*

Tadpoles, *26*
Turtles, 32, *33*

Walruses, 44
Water moccasin, *34*
Whales, 45
Worms, 15

About the Author

MARGARET MERRILL's interest in sharing her love of science with children is the basis for this book, her first for young readers. She has worked with a natural science program at the Vermont Institute of Natural Science, and has directed a natural science camp, the Nineveh Day Camp, in Vermont.

Ms. Merrill lives in Woodstock, Vermont, where she enjoys exploring the country in all its seasons with her husband and their two sons.

About the Artist

PAMELA CARROLL has illustrated several science books for children, and has done extensive illustrations for publications such as *Reader's Digest* and *The New York Times*. She has also taught art to grade school children.

Ms. Carroll received her education at the Académie Julian, in Paris, and at the Pratt Institute, New York. She lives with her husband and their two children in Westmoreland, New Hampshire.